Geometry
(4-6)

© 2015 OnBoard Academics, Inc
Portsmouth, NH
800-596-3175
www.onboardacademics.com
ISBN: 978-1-63096-082-7

OnBoard Academic's books are specifically designed to be used as printed workbooks or as on-screen instruction. Each page offers focused exercises and students quickly master topics with enough proficiency to move on to the next level.

OnBoard Academic's lessons are used in over 25,000 classrooms to rave reviews. Our lessons are aligned to the most recent governmental standards and are updated from time to time as standards change. Correlation documents are located on our website. Our lessons are created, edited and evaluated by educators to ensure top quality and real life success.

Interactive lessons for digital whiteboards, mobile devices, and PCs are available at www.onboardacademics.com. These interactive lessons make great additions to our books.

You can always reach us at customerservice@onboardacademics.com.

Angles

Key Vocabulary

acute angle

obtuse angle

right angle

Describe a right angle.

A right angle is an angle that has a measure of 90°

90°

List 5 things around you that have right angles.

1._____

2._____

3_____

4._____

5._____

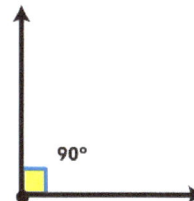

Here are some other types of angles.

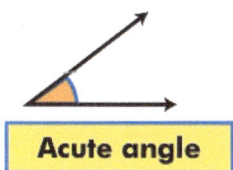

Acute angle

An angle with a measure of less than 90°.

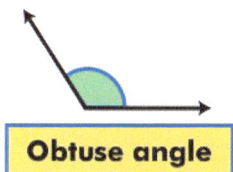

Obtuse angle

An angle with a measure of greater than 90°, but less than 180°.

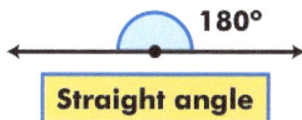

180°

Straight angle

An angle with a measure of 180°.

Label these angles.

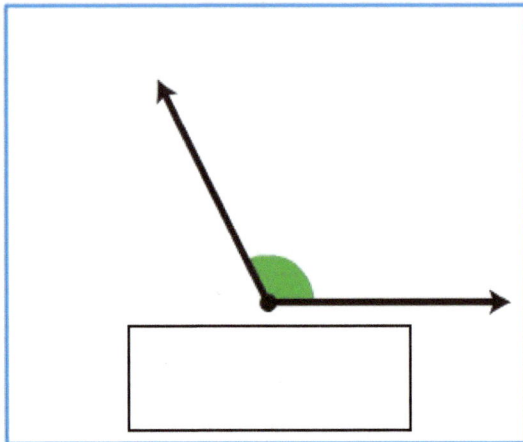

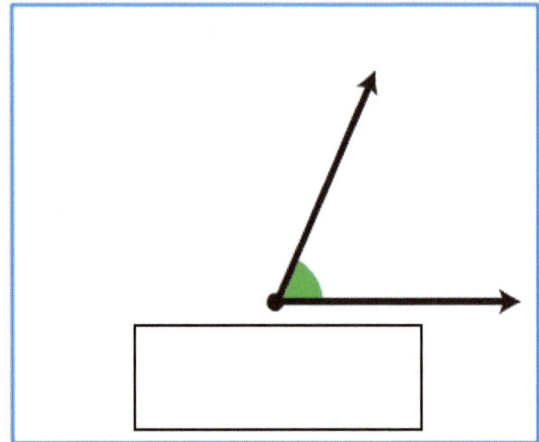

Label these angles.
Use the letters in the circles for your labels.

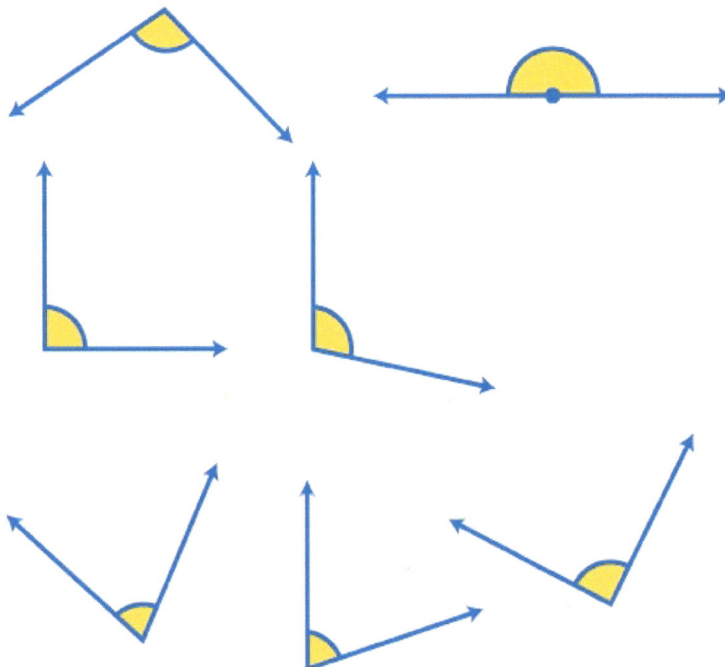

R ight angle

A cute angle

O btuse angle

S traight angle

Estimate the measure of these angles.

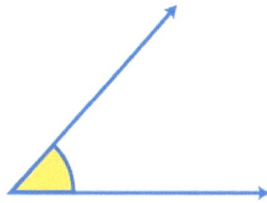 

| Estimate | Estimate |

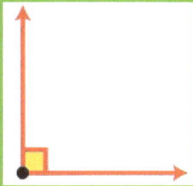 A right angle gives you the information that you need to estimate the measurement.

USING A PROTRACTOR

Make sure that you position the protractor correctly (Figure 1).

Always read up from zero using either the outside scale (Figure 2) or the inside scale (Figure 3).

Read the protractor to determine the measure of the angle (Figure 4).

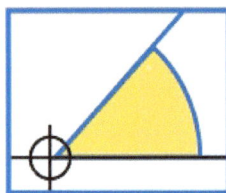

Figure 1

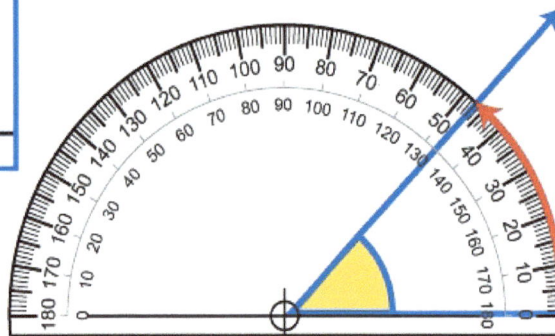

Figure 2

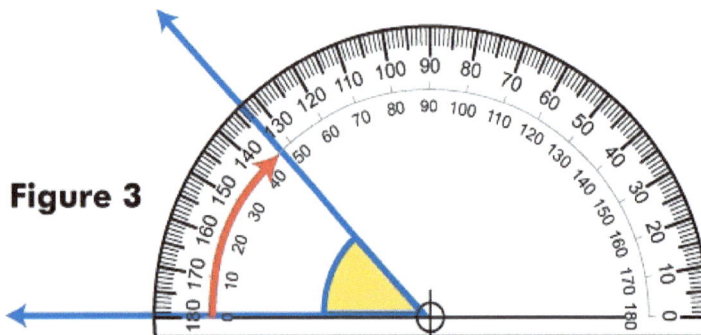

Figure 3

Figure 4

Name_____

Angles Quiz

Fill in or circle the correct answer.

1 True or false? An obtuse angle is greater than a right angle.

2 What is the measure of a straight angle?

- **A** 90°
- **B** 45°
- **C** 100°
- **D** 180°

3 What is the measure of the obtuse angle (in degrees)?

30°

4 What is the measure of the acute angle (in degrees)?

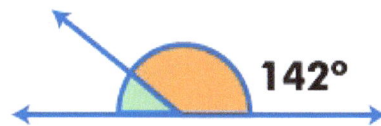

142°

Angle Properties

Key Vocabulary

right angle

obtuse angle

acute angle

supplementary angles

protractor

Draw a line to connect the angles with its correct angle facts.

| Right | Acute | Obtuse |

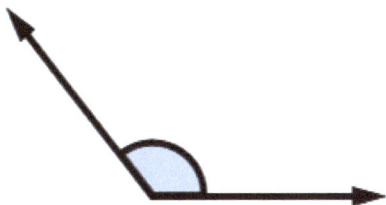

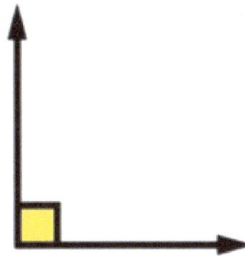

| Less than 90° | Greater than 90° & less than 180° | 90° |

Angle Estimation Game

This game can be played in teams or individually. If completing individually enter your answers into the Team 1 line.

OVERVIEW
1. Divide the class into two teams: Team 1 and Team 2.
2. Ask each Team to estimate the measure of each of the 3 angles.
3. Record the estimates in the left side of the table cells. You can record the score that the Team receives for each estimate in the right side of the table cells (see Scoring below).

SCORING
Award 5 points for an estimate that is within 5° of the correct answer.
Award 2 points for an estimate that is within 10° of the correct answer.
Award 0 points for an estimate that is incorrect by more than 10°.
Deduct 2 points for an estimate that is incorrect by more than 20°.
Sum the points to determine the winning Team.

MEASURING THE ANGLES
When the Teams have made their estimates, copy and paste the protractor from this page to measure the angles. Draw a rectangle using the Magic Pen Tool to zoom in on the measurement.

Note that the angle images are locked and so students will not be able to rotate them to make their estimates, but they can rotate the protractor to measure the angles.

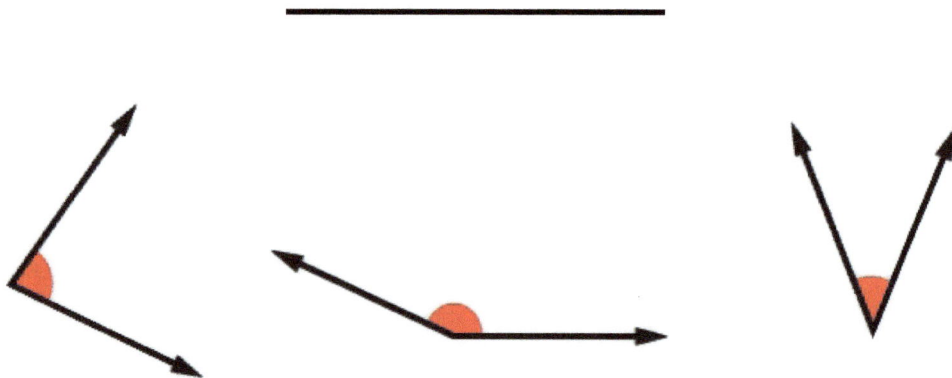

	Angle 1	Angle 2	Angle 3	Total
Team ①	°	°	°	
Team ②	°	°	°	

The sum of the measure of the angles on a straight line is 180⁰

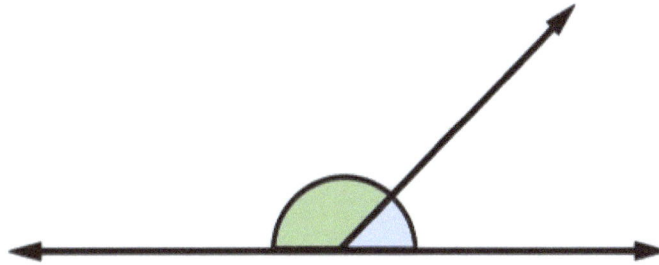

180°   **If the sum of two angles is 180°, then the angles are said to be *supplementary*.**

_____

Match the angles and the segments.

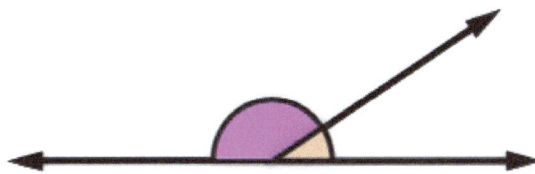

147°

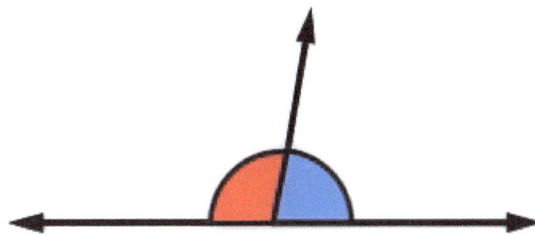

100°

45°

135°

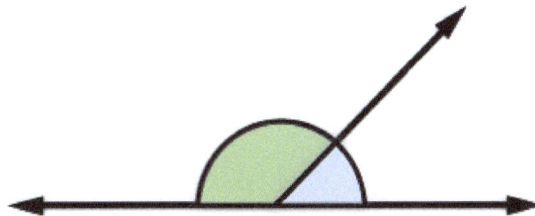

80°

What is the measure of the missing angle?

Name: _____

Angle Properties Quiz

1 **Are these angles correctly classified?**

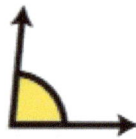  

right   obtuse   acute

2 **What is the measure of angle n?**

- **A** 16°
- **B** 196°
- **C** 14°
- **D** 106°

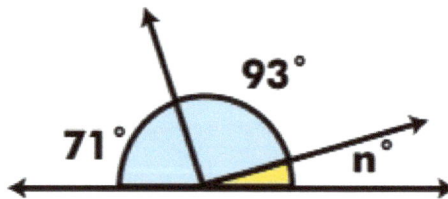

3 **What is the measure of angle a (in degrees)?**

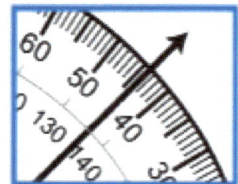
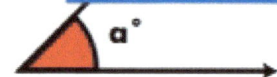

The Coordinate Plane

Key Vocabulary

coordinate plane

quadrant

origin

ordered pair

Find the ordered pairs.

Hint

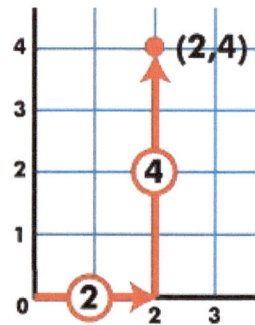

Washington, D.C. Downtown Area

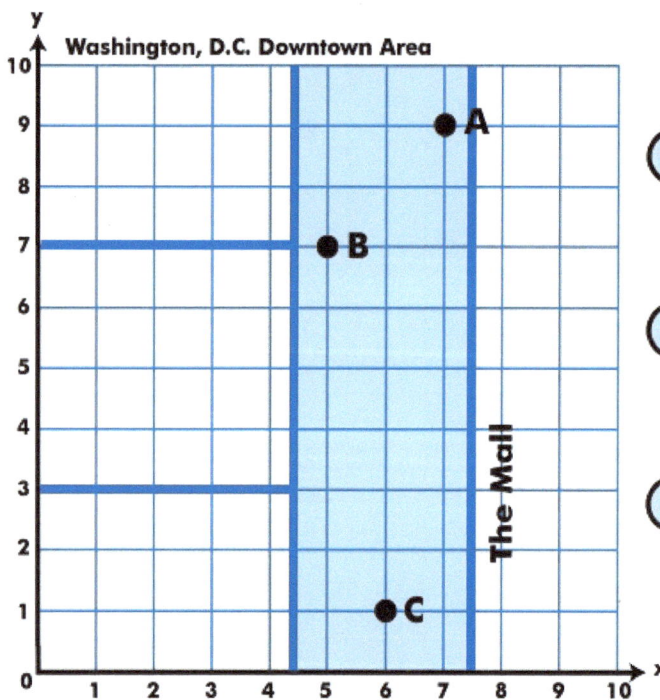

The Mall

(A) **Air & Space Museum**

$$(\underline{\quad} , \underline{\quad})$$

(B) **American History Museur**

$$(\underline{\quad} , \underline{\quad})$$

(C) **Lincoln Memorial**

$$(\underline{\quad} , \underline{\quad})$$

Plot the ordered pairs.

Can you guess which monuments are (6,6) and (2,5)?

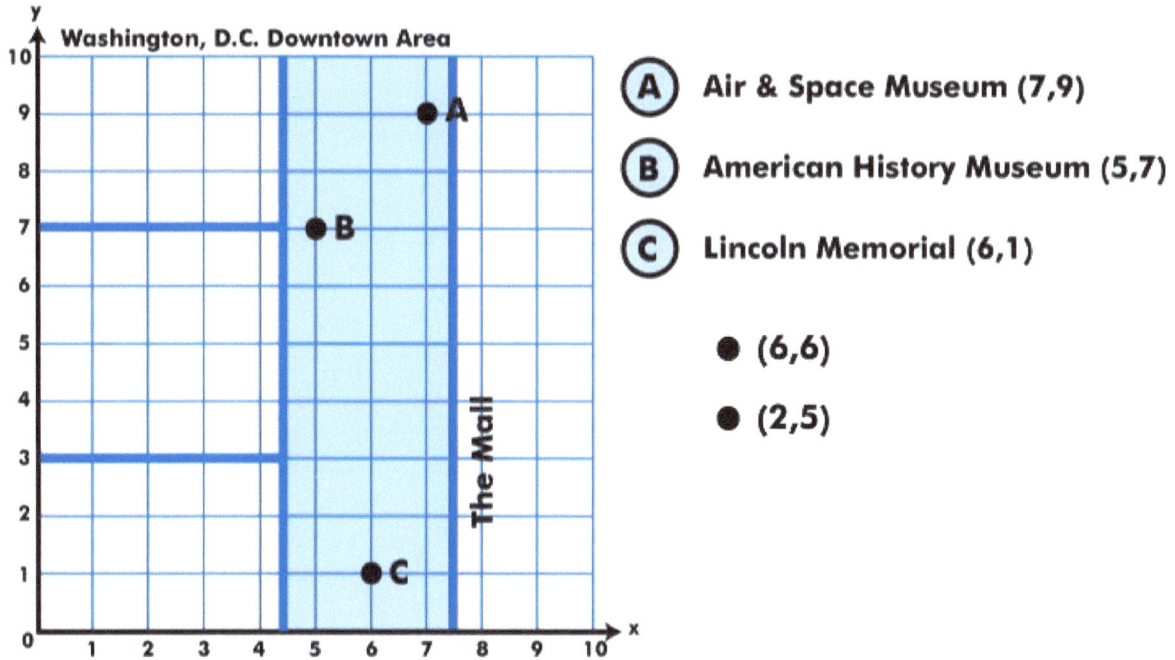

(A) Air & Space Museum (7,9)

(B) American History Museum (5,7)

(C) Lincoln Memorial (6,1)

● (6,6)

● (2,5)

(6,6) is the Washington Monument

(2,5) is the White House

Find the ordered pairs.

<u>Hint</u>

(3,-4)

_____

WASHINGTON D.C.

A	Air & Space Museum
B	American History Museum
C	Washington Monument
D	The White House
E	Lincoln Memorial
F	Watergate Bldg.
G	Dupont Circle
H	Arlington National Cemetry

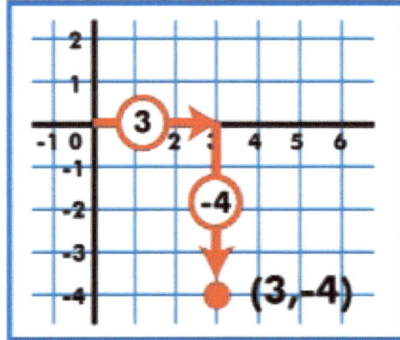

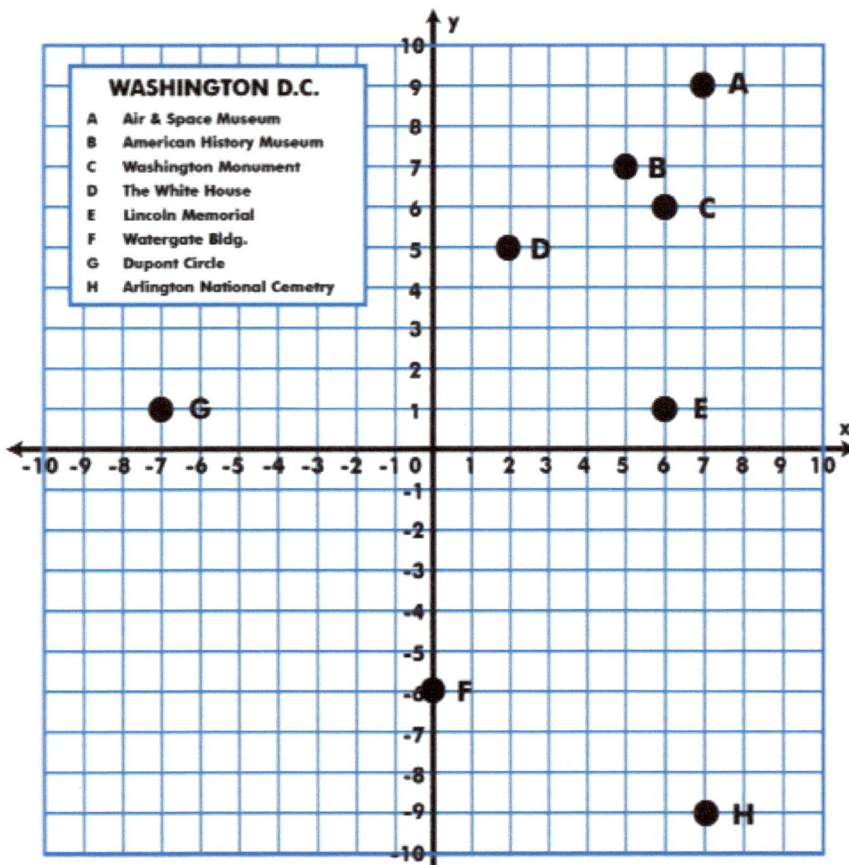

(F) **Watergate Building**

(__ , __)

(G) **Dupont Circle**

(__ , __)

(H) **Arlington Cemetry**

(__ , __)

Circle the image with the quadrants labeled correctly?

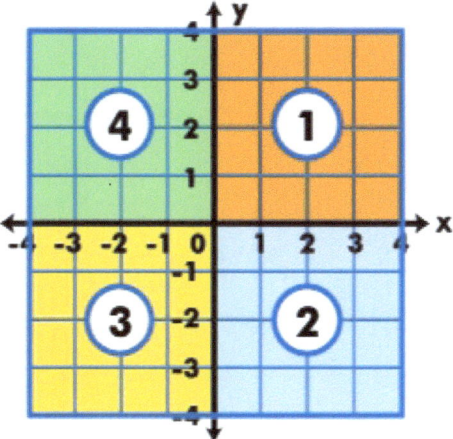

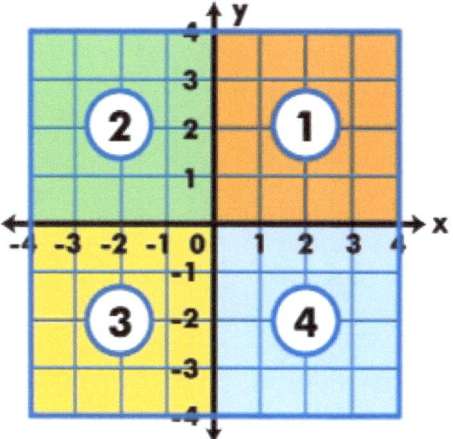

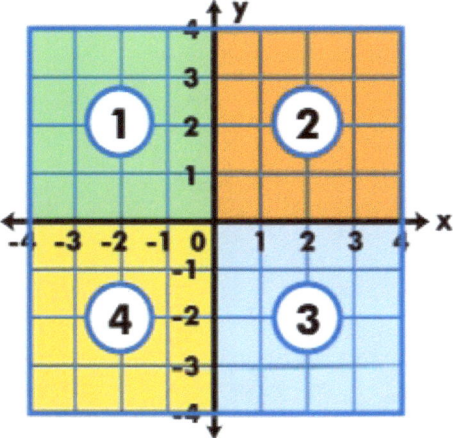

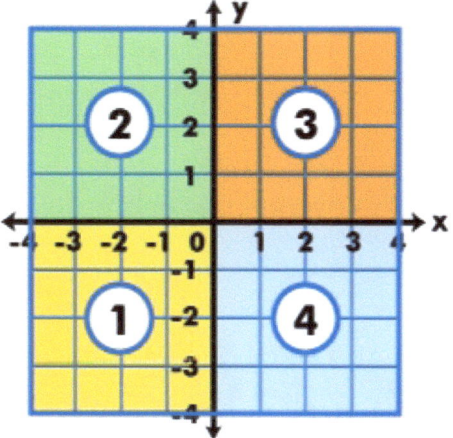

Write the ordered pair for A, B, and C.

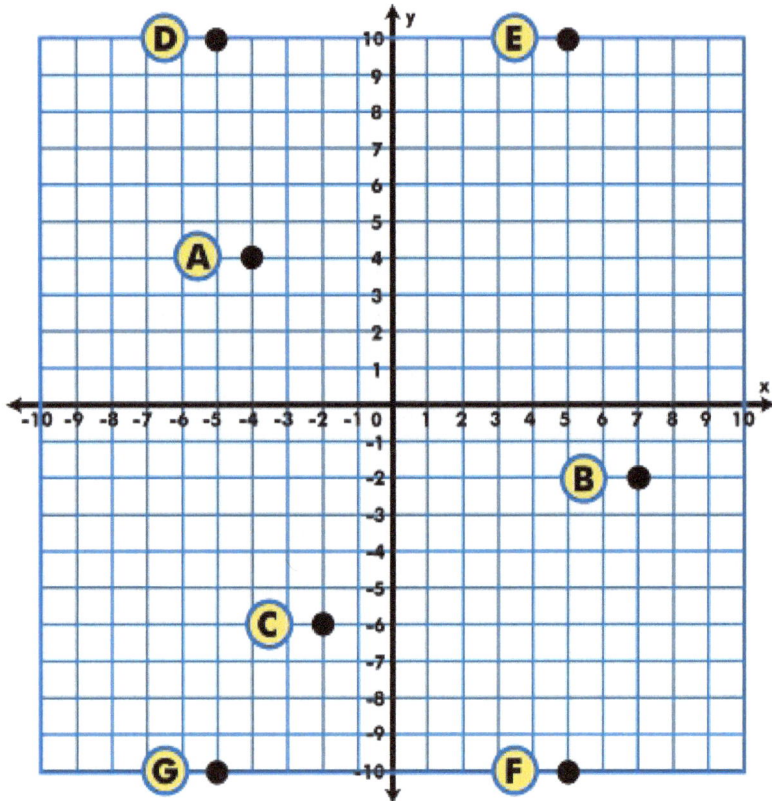

A (___ , ___)

B (___ , ___)

C (___ , ___)

Which point has the coordinates (-5,10)?

Plot the points for A, B and C and then find an ordered pair for D to form a parallelogram.

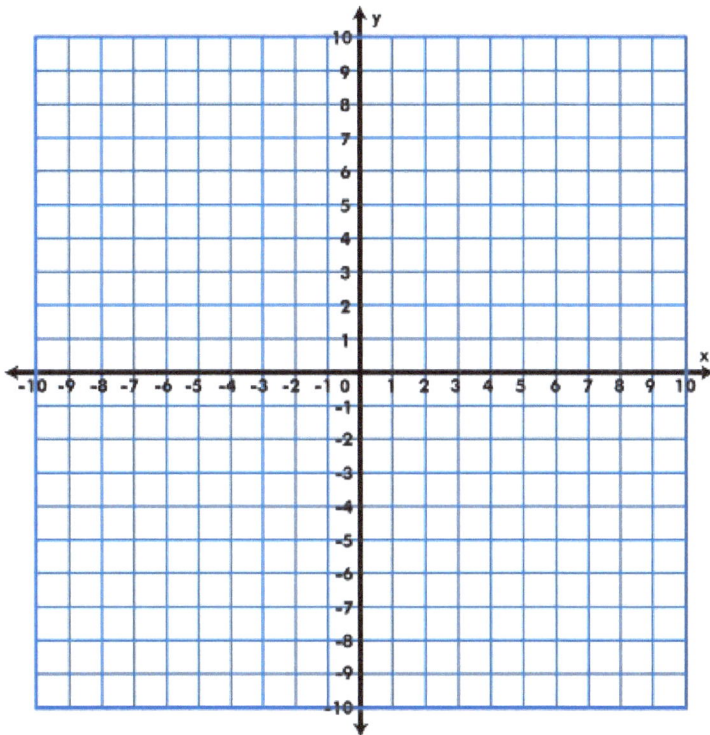

A (-5,-4)   ●

B (5,-4)   ●

C (0,2)   ●

D (__ , __)   ●

Graphing Functions on the Coordinate Plane
Draw the points on the coordinate plane.

$y = 2x + 2$

x	y
1	4
2	6
3	8
4	10

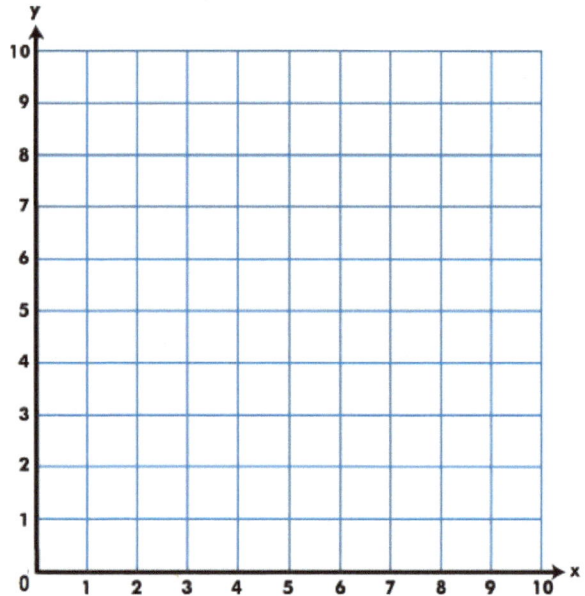

Complete the graph for y = 3x + 1

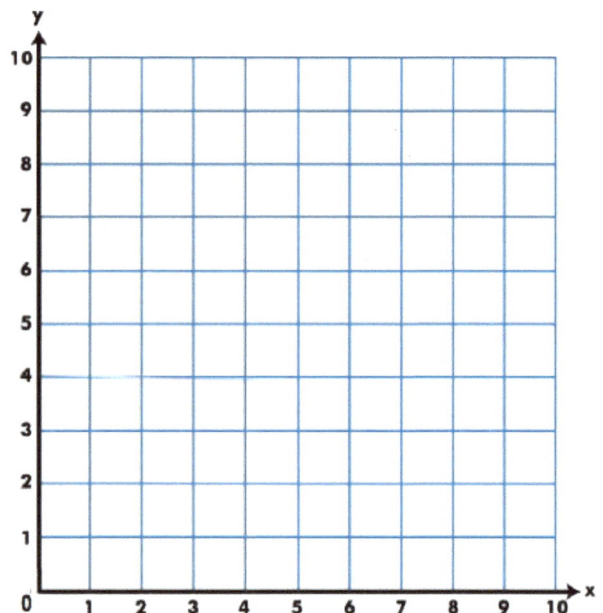

Name: _____

The Coordinate Plane Quiz

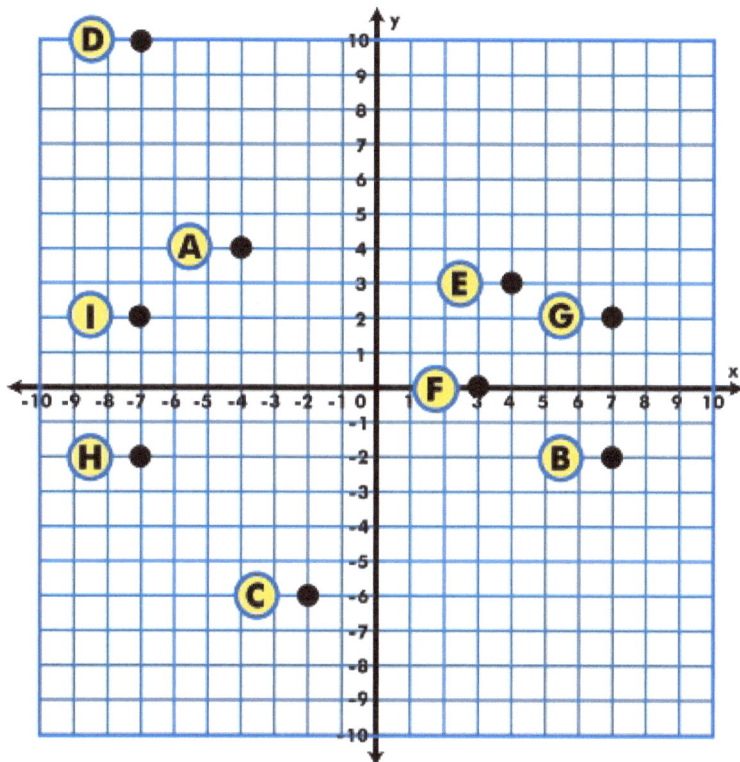

1. Which point has the coordinates (7,-2)?

2. Which point has the coordinates (-7,2)?

3. What is the y-coordinate of point F?

4. What is the y-coordinate of point E?

5. What is the x-coordinate of point E?

The Coordinate Plane

Key Vocabulary

coordinate plane

ordered pair

quadrant

x-axis

Study the graph and its labels below.
Note the x and y axis, the quadrants and the origin.

Ordered

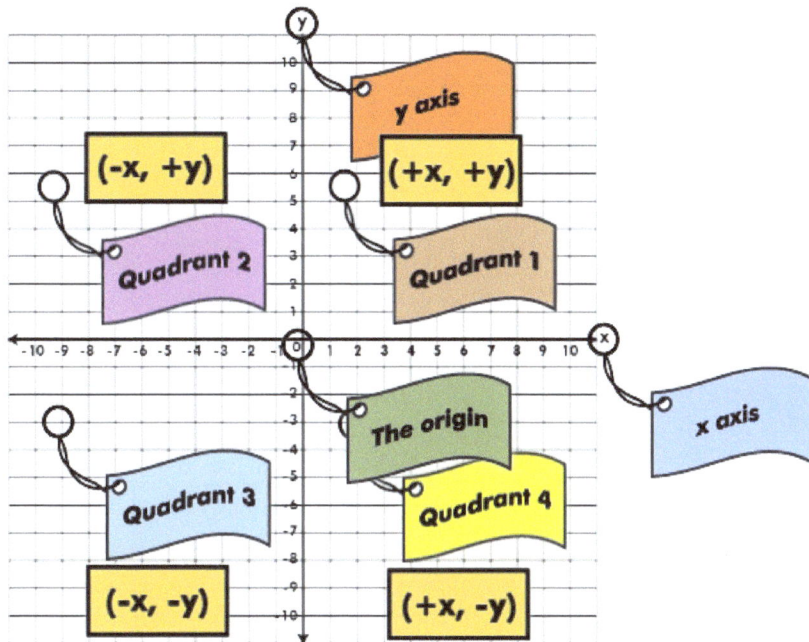

Pairs

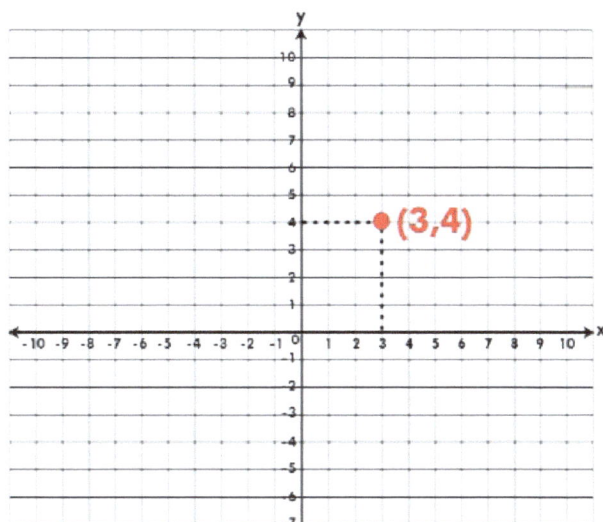

(3,4) is the ordered pair. The x-coordinate is the first coordinate in an ordered pair and the y-coordinate is second.

Write the ordered pairs for A, B and C.

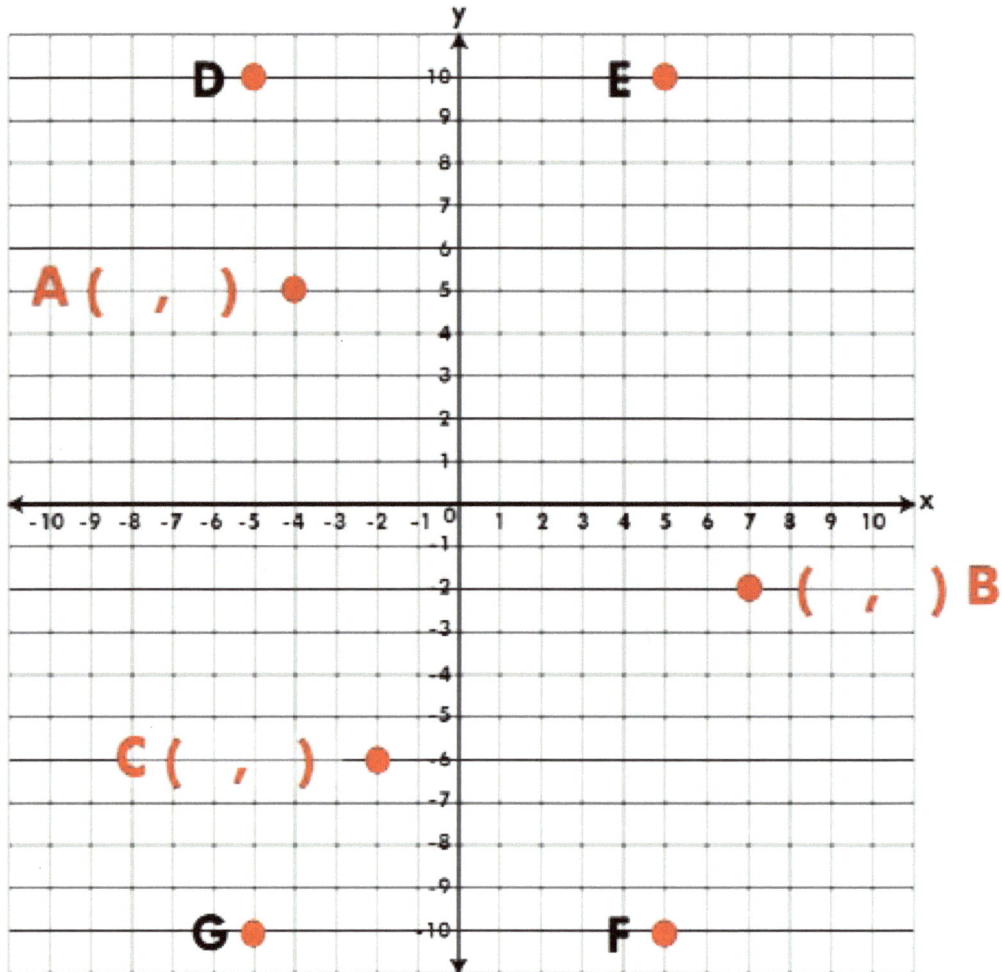

Which point has the coordinates (-5,10)?

Add and label a point to form an isosceles triangle.

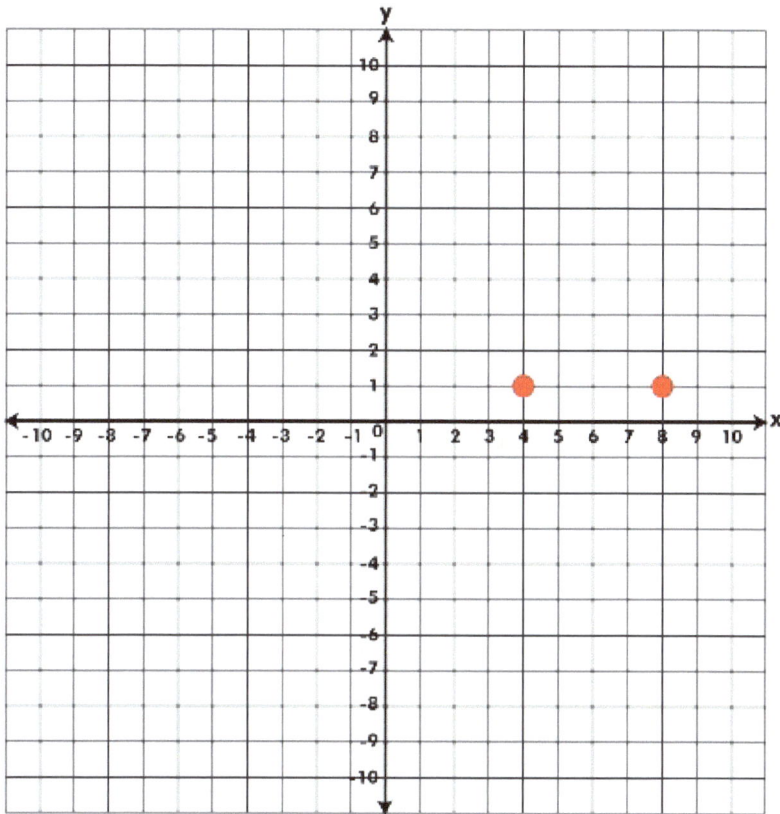

In which quadrant(s) do the points lie?

Graph the points.

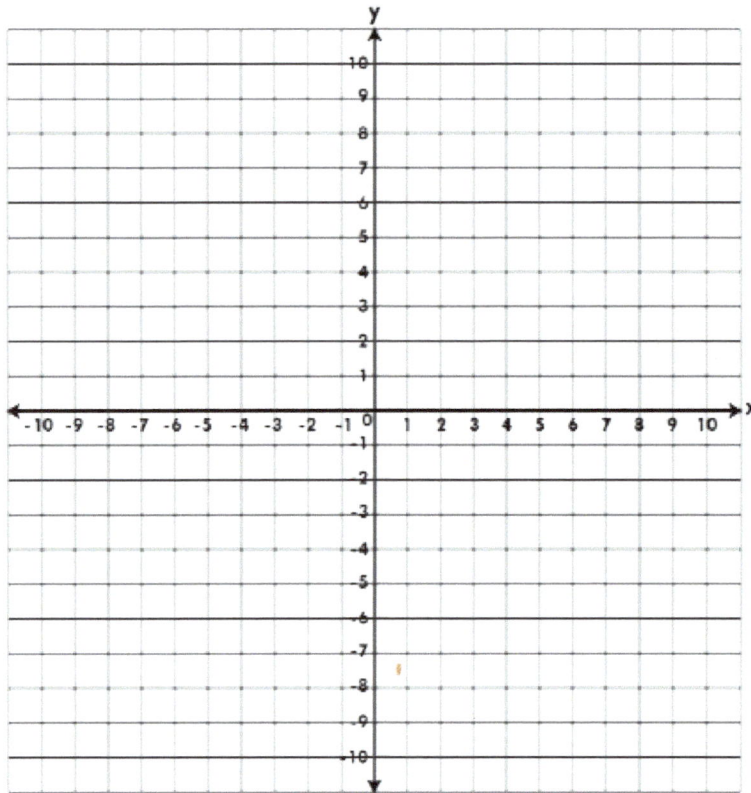

A (4,0)

(0,4) B

(0,0) C

D (4,4)

If you connect the points, what shape have you made?

Find the missing point for a hexagon.

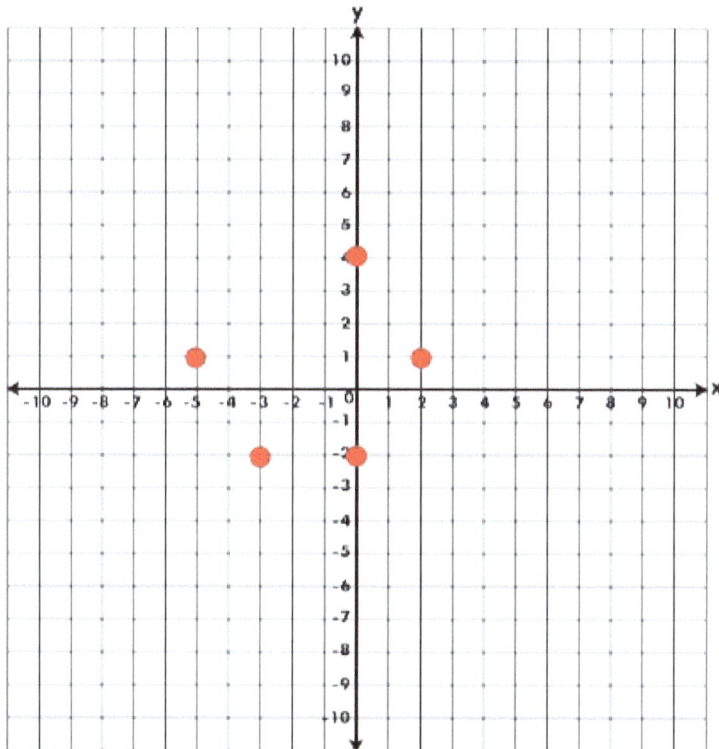

What is the point that will complete a hexagon?

(___ , ___) •

In which quadrant does the point lie?

Name_____

The Coordinate Plane Quiz

1 True or false: the ordered pair for G is (-5,5)

2 Which of the ordered pairs are located in quadrant 2?

A (-4,1) and (6,-3)

B (4,-1) and (-6,3)

C (-4,-1) and (-6,-3)

D (-4,1) and (-6,3)

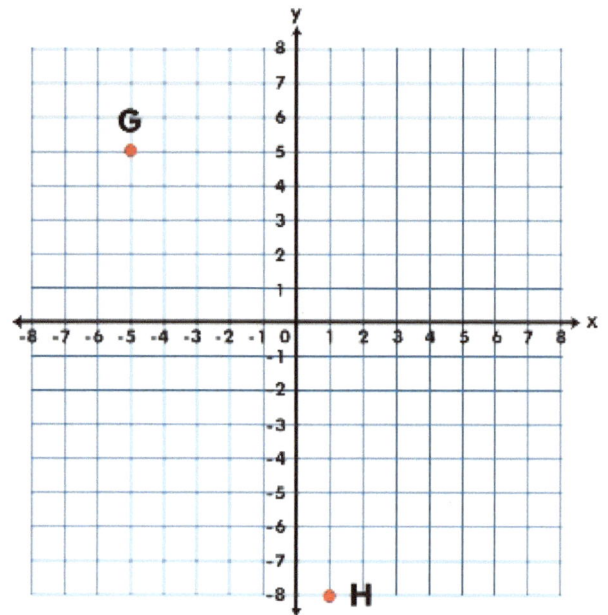

3 If FGH forms an isosceles triangle, what is the x coordinate of F?

4 If FGH forms an isosceles triangle, what is the y coordinate of F?

Transformations

Key Vocabulary

rotate

reflect

translate

congruent

mirror line

transformation

Are these shapes congruent?

Congruent: figures that are
the same shape and size.

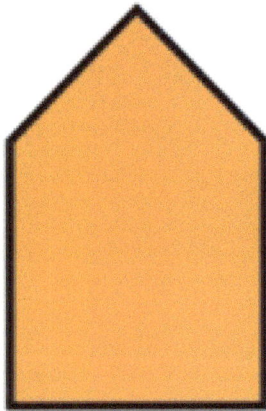

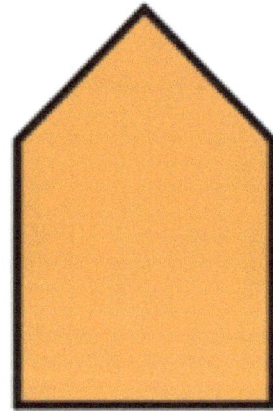

How can you be sure that these shapes are congruent or not? _____

_____

_____

_____

Describe the movement to superimpose one figure over another.

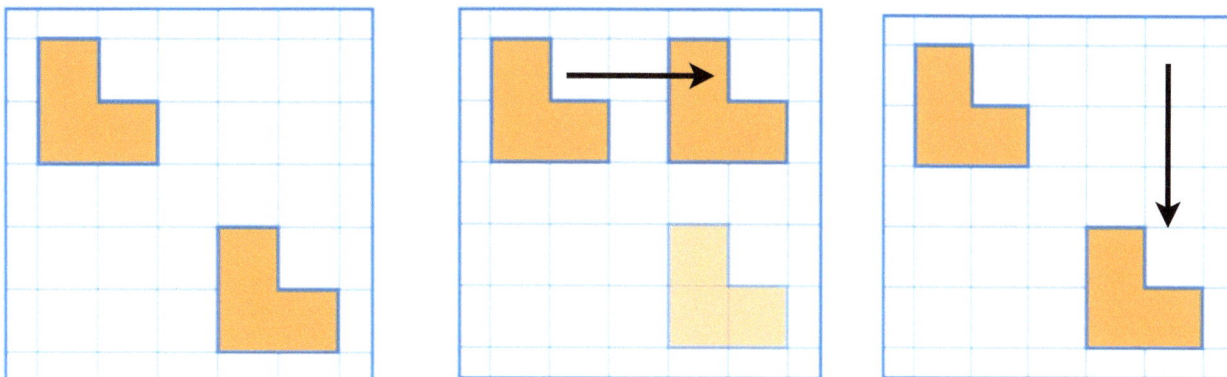

Step 1 Move _____ spaces to the _____.

Step 2 Move _____.

We call this *a translation or a slide.*

We call this *a reflection or a flip.*

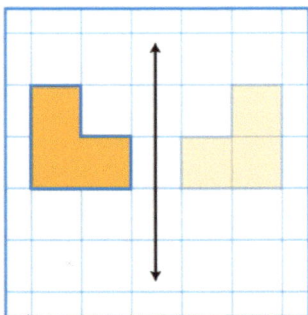

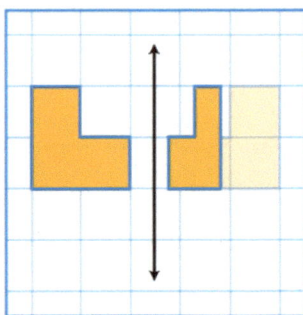

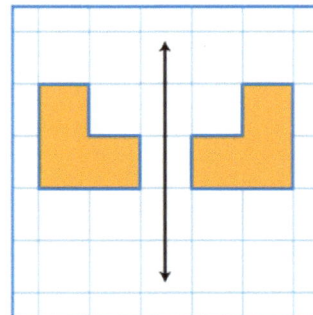

Think about turning a page in a book. Is this a reflection or a flip? _____

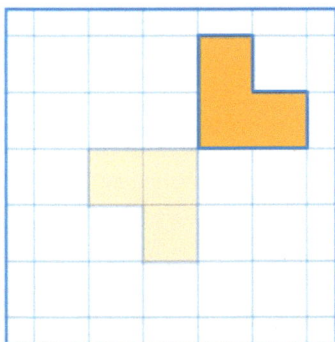

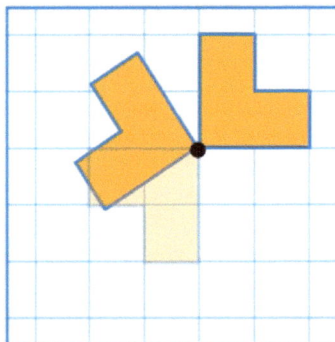

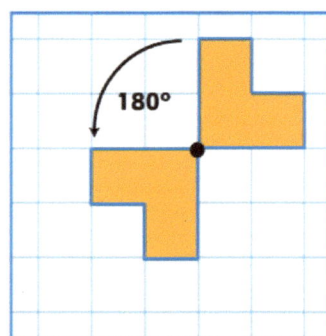

_____

We call this *a rotation or a turn.* Drag
the point to the center of the rotation.

Show and describe a translation.

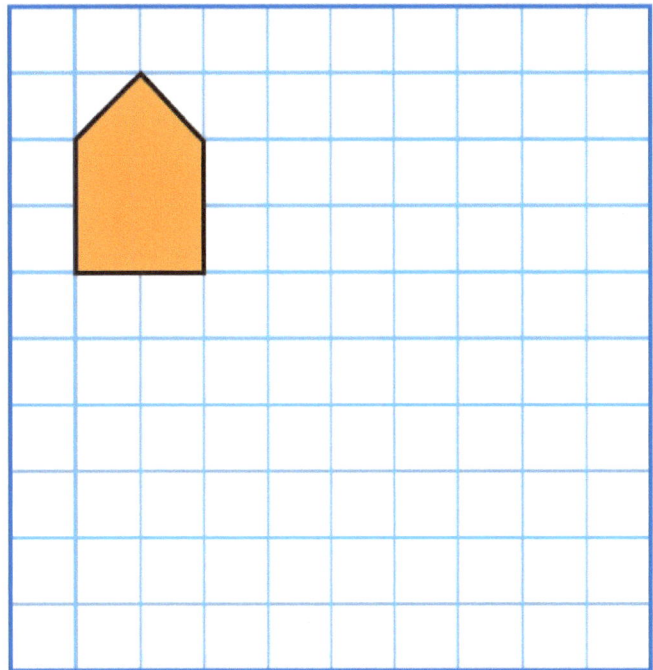

_____

_____

_____

_____

_____

_____

_____

Draw a reflection.

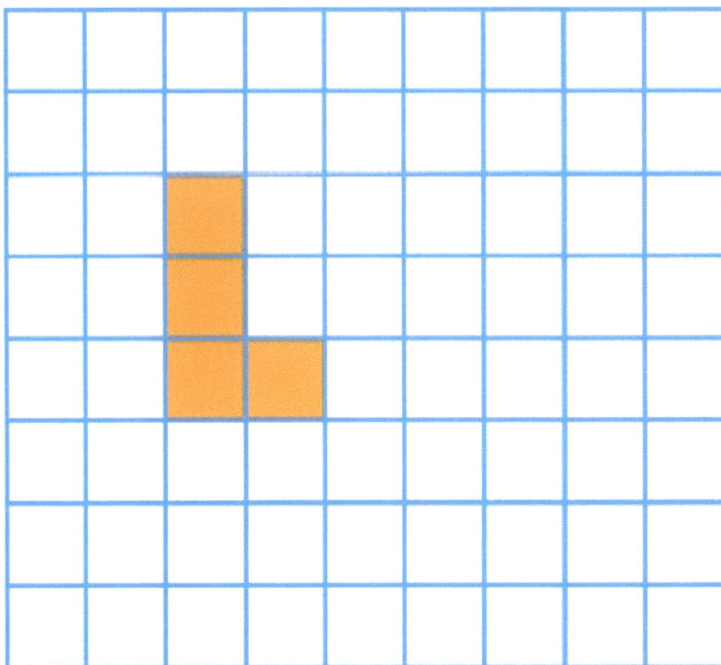

Can you show a rotation using this graph. Use the point as the center for the rotation.

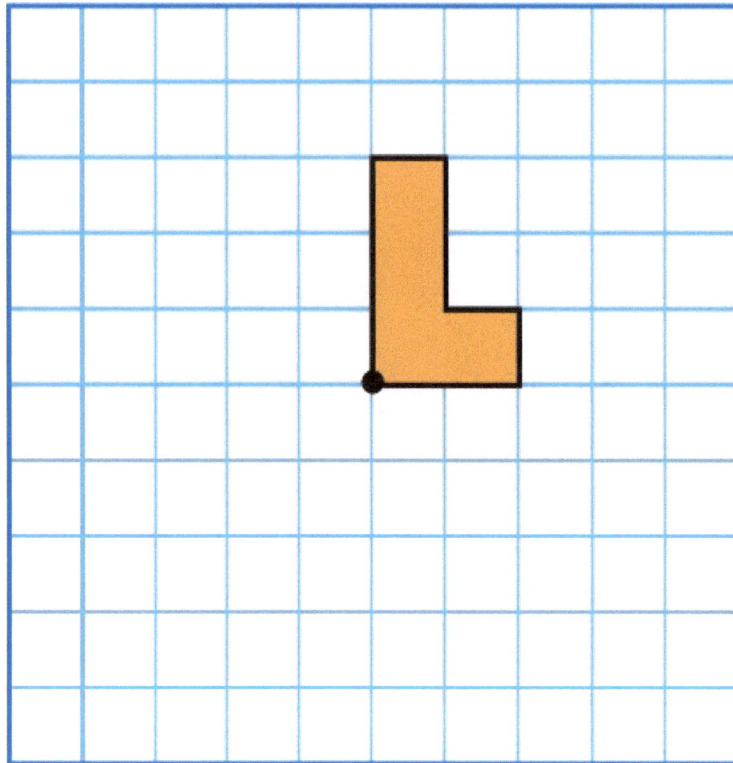

Label the transformations.

How many degrees.

How many degrees did each figure rotate when moving clockwise and counter-clockwise. The yellow figure is the starting point.

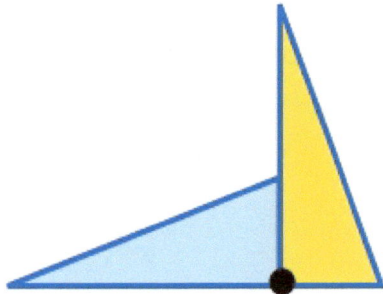

Clockwise _____

Counter Clockwise _____

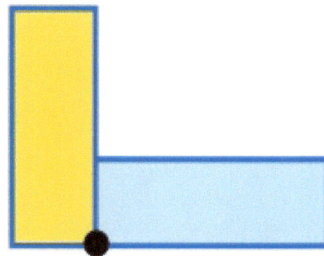

Clockwise _____

Counter Clockwise _____

Name_____

Transformation Quiz

Answer yes or no in the gold box for each figure and label.

Decide if each transformation has been correctly described.

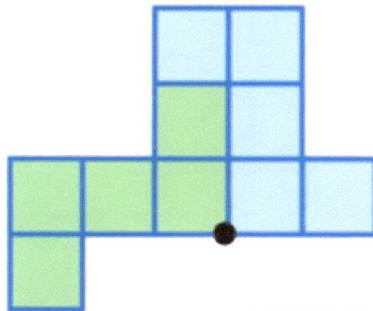

Rotation

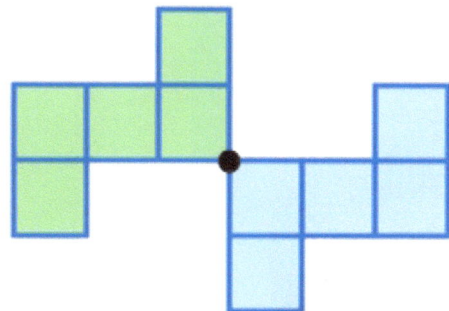

Rotation

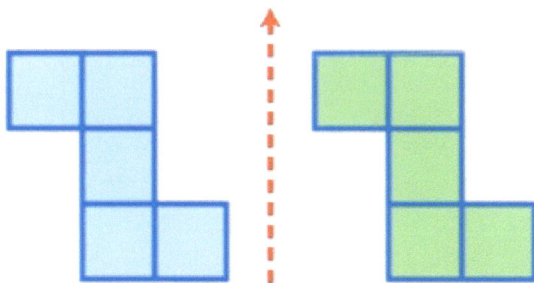

Reflection

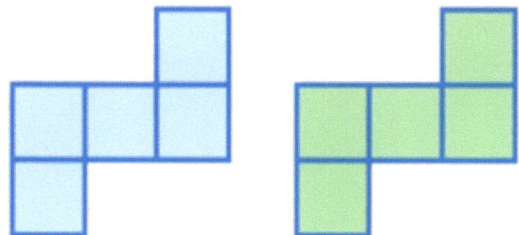

Translation

Transformations of the Coordinate Plane

Key Vocabulary

transformation

rotate

reflect

translate

coordinate plane

Translate triangle A per the instructions below.
Draw the triangle in the two positions as instructed. The first one is completed for you.

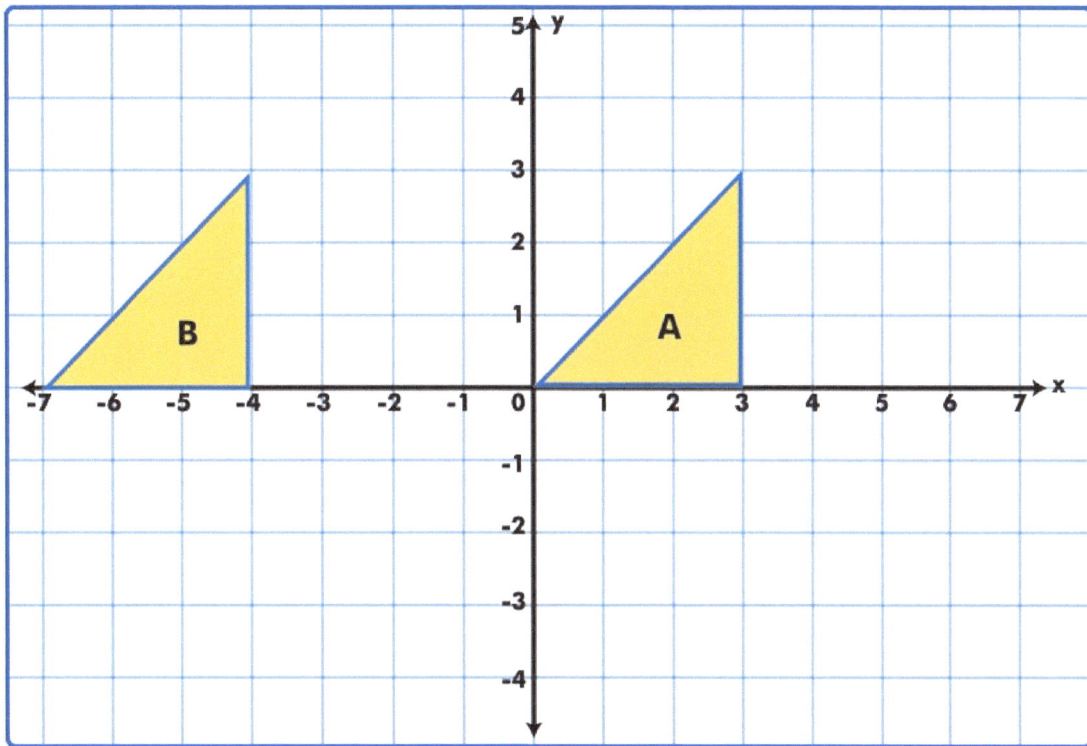

Translate Triangle A
7 left and label as B.

Translate Triangle A
4 right and 1 up and
label as C.

Translate Triangle A
3 right and 4 down
and label as D.

Shape A is reflected across the y-axis and labeled as B.

Study the illustration and then for the second illustration reflect shape A across the y-axis.

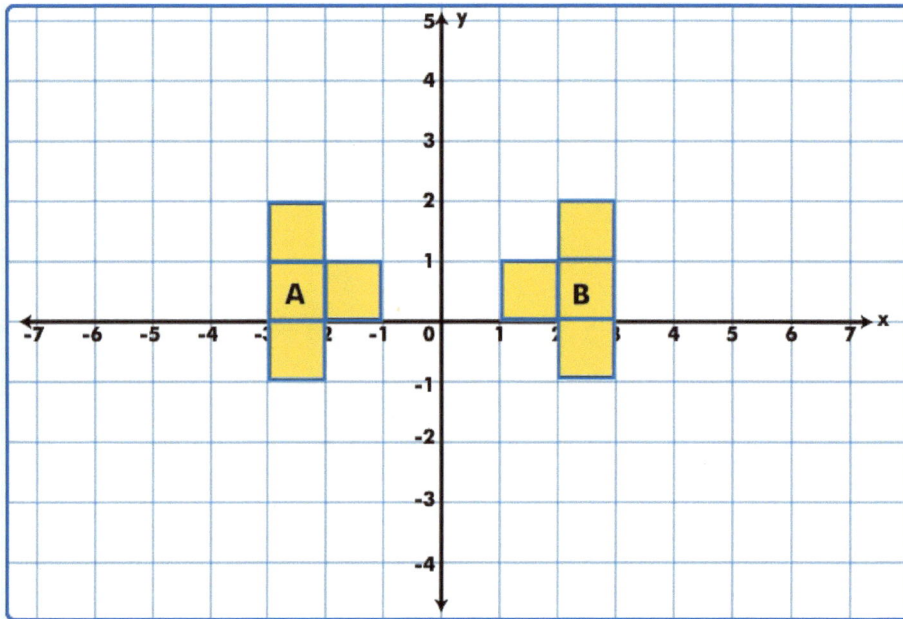

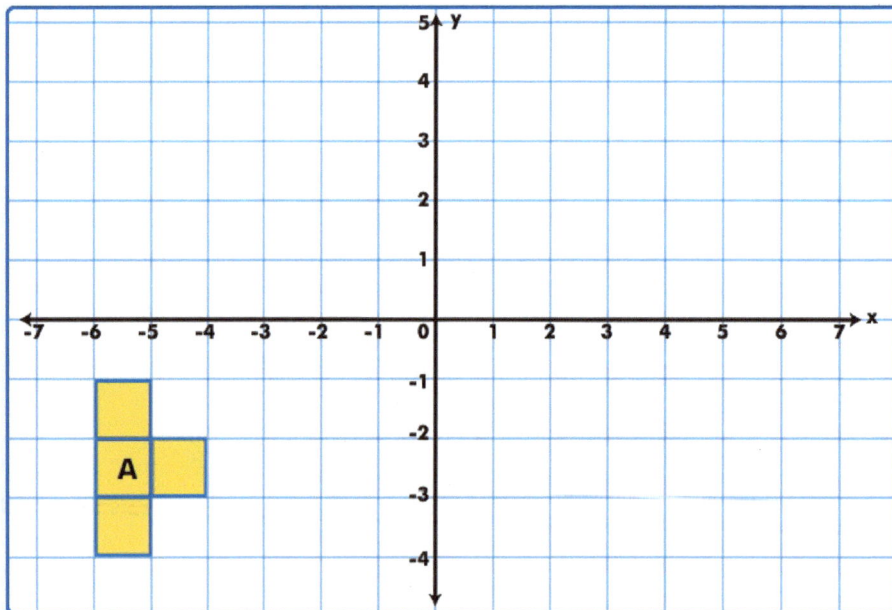

Reflect shape A across the x-axis and label it B.

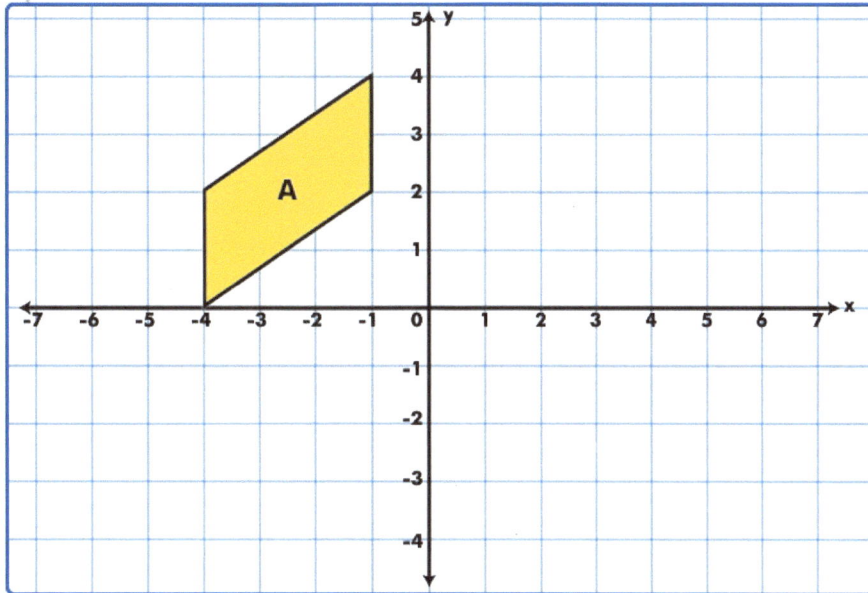

Reflect and label points A, B and C across the mirror line.

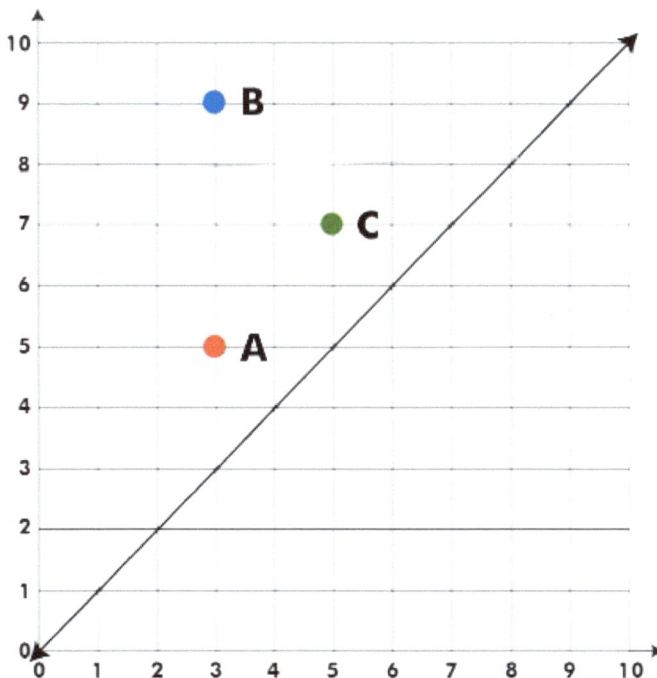

A' (____ , ____)

B' (____ , ____)

C' (____ , ____)

Name_____

Transformations of the Coordinate Plane Quiz

1 True or false? If any point (x,y) is reflected across this mirror line, the coordinate pair of the reflected point will be (y,x).

2 Which image is a translation of trapezoid E?

A A

B B

C C

D D

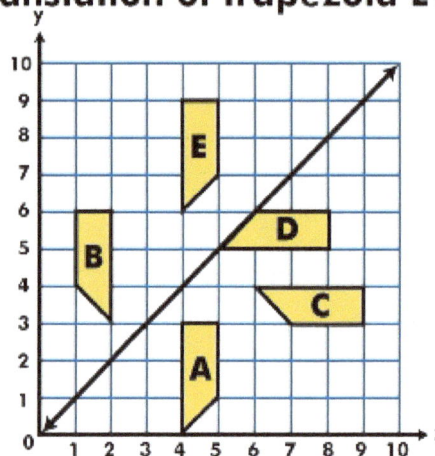

3 The translation that mapped E onto A was ___ units in the horizontal direction?

4 The translation that mapped E onto A was ___ unit in the vertical direction?

Properties of Quadrilaterals

Key Vocabulary

quadrilateral

parallelogram

trapezoid

rhombus

congruent

adjacent

Quadrilaterals

Polygon: a plane figure with at least three straight sides and angles.

Quadrilateral: a closed four-sided figure having four straight sides

A quadrilateral is a polygon with ⬚ **sides and** ⬚ **angles**

The sum of the measure of the angles in a quadrilateral is ⬚°

Classifying Quadrilaterals: Parallelograms.

Step 1

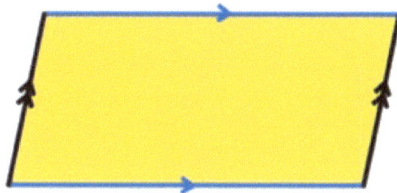

- **Two pairs of parallel sides**

Step 2

10 in

5 in   5 in

10 in

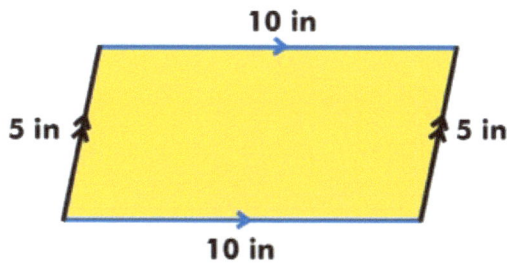

- **Two pairs of parallel sides**
- **Opposite sides are congruent**

Step 3

10 in

5 in   5 in

10 in

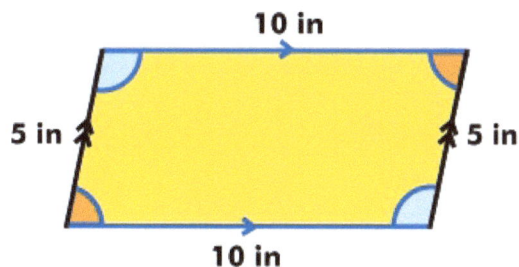

- **Two pairs of parallel sides**
- **Opposite sides are congruent**
- **Opposite angles are congruent**

Congruent: Same size and shape

Label the quadrilaterals and them sort them according to if they area also a parallelogram.

Use the three step classifying steps from the previous page.

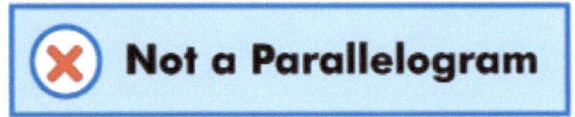

✓ **Parallelogram**	✗ **Not a Parallelogram**

trapezoid	kite	rhombus	rectangle	square

Square: a plane figure with four equal straight sides and four right angles

Rectangle: a plane figure with four straight sides and four right angles

Parallelogram: a plane figure with opposite sides being parallel

Rhombus: a parallelogram with opposite equal acute angles, opposite equal obtuse angles, and four equal sides.

Trapezoid: a quadrilateral with only one pair of parallel sides

Classify these quadrilaterals.

S quare

R ectangle

P arallelogram

R hombus

T rapezoid

Name: _____

Properties of Quadrilaterals Quiz

1 True or false? A trapezoid is a parallelogram.

2 A square can be classified as which of the following?

A Parallelogram

B Rhombus

C Rectangle

D All of the above

3 True or false? A rectangle is always a parallelogram, but a parallelogram is not always a rectangle.

4 Three angles of a quadrilateral have the measures of 83°, 130°, and 95°. What is the measure of the 4th angle?

Triangle Classification

Key Vocabulary

equilateral triangle

isosceles triangle

acute angle

obtuse angle

right angle

What do you think the measurement of the third side of this triangle is? _____

An equilateral triangle

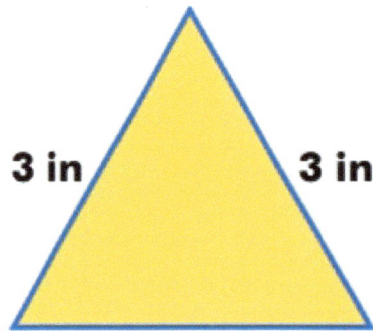

3 in   3 in

isosceles triangle

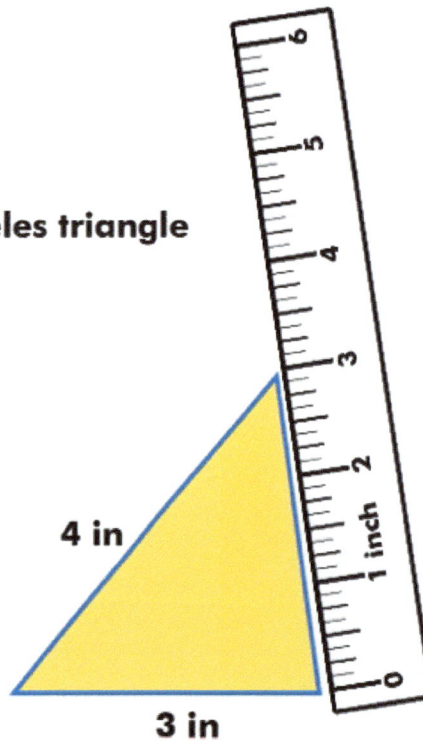

4 in

3 in

How is an isosceles triangle different from an equilateral triangle? _____

_____.

Draw a triangle with no congruent sides.
This is called a **scalene triangle**.

Connect the triangle facts with the correct triangle.

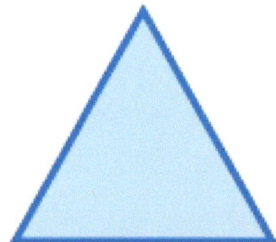

Equilateral

Isosceles

Scalene

| Two angles of equal measure |
| No sides of equal length |

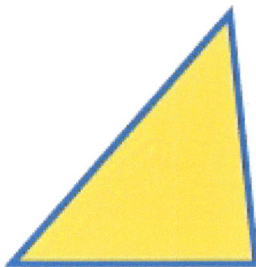

| Three sides of equal length |
| No angles of equal measure |

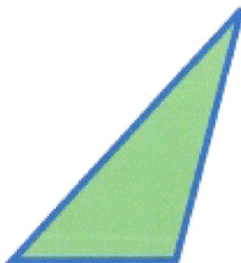

| Three angles of equal measure |
| Two sides of equal length |

Classify these triangles by their sides.

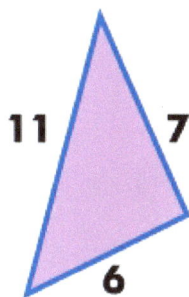

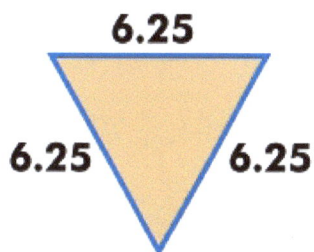

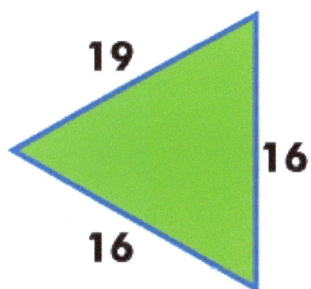

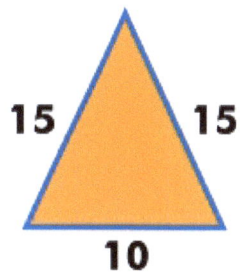

11   7

6

17   19

4

6.25

6.25   6.25

19

16

16

15   15

10

(E) quilateral

(I) sosceles

(S) calene

We can also classify triangles by the measure of their angles.

Right triangle

One angle is a right angle (90˚)

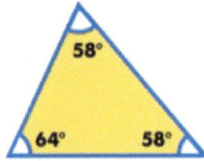

58°
64°   58°
Acute Triangle

All three angles are acute (< 90˚)

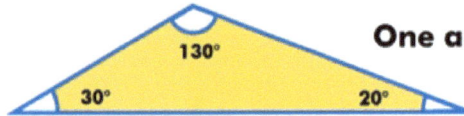

130°
30°   20°
Obtuse triangle

One angle is obtuse (> 90˚)

_____

Draw these triangles.

Scalene right

Equilateral acute

Isosceles obtuse

Equilateral right

Name: _____

Triangle Classification Quiz

1 True or false? Every equilateral triangle is an isosceles triangle.

2 True or false? Every isosceles triangle is an equilateral triangle.

3 Classify Figure 1.

- **A** Isosceles acute
- **B** Equilateral acute
- **C** Scalene acute
- **D** Scalene obtuse

4 Classify Figure 2.

- **A** Isosceles acute
- **B** Equilateral acute
- **C** Scalene acute
- **D** Scalene obtuse

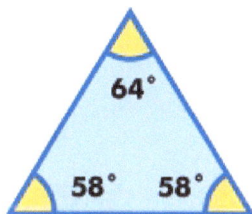

Figure 1

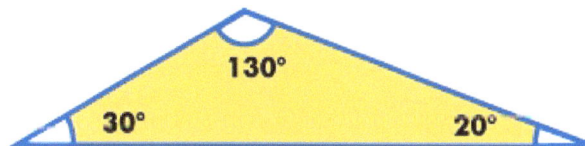

Figure 2

Exploring Symmetry

Key Vocabulary

line symmetry

rotational symmetry

congruent/congruence

Draw the lines of symmetry for this rectangle.

This figure has [ ] lines of symmetry.

A line of symmetry is a line that divides a figure into two congruent parts, each of which is the mirror image of the other.

Study the triangles' lines of symmetry and label the triangle types.

1

0

3

Jesús was asked to draw lines of symmetry for these quadrilaterals. Circle the ones drawn correctly.

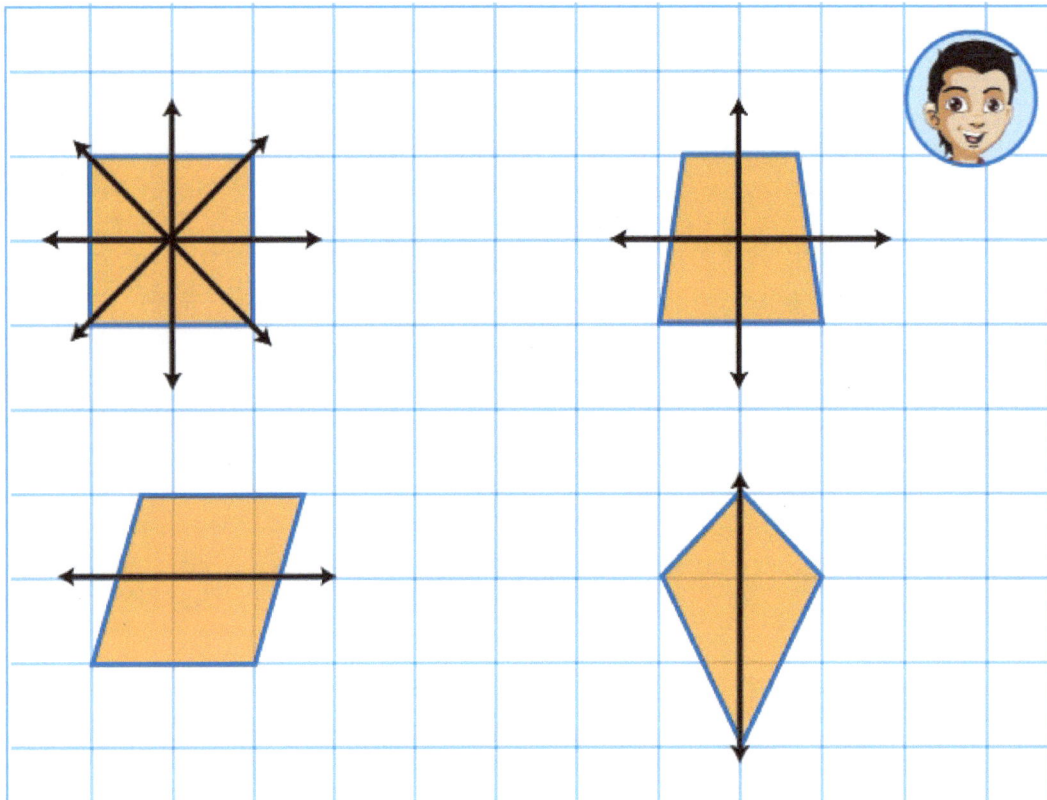

Rotational Symmetry

A figure has rotational symmetry if it looks the same at any point when it is rotated less than a full turn.

The top of the triangle is noted by a red dot. Notice the shape as it rotates counter clockwise. Circle the positions of rotational symmetry.

The equilateral triangle has rotational symmetry at these angles.

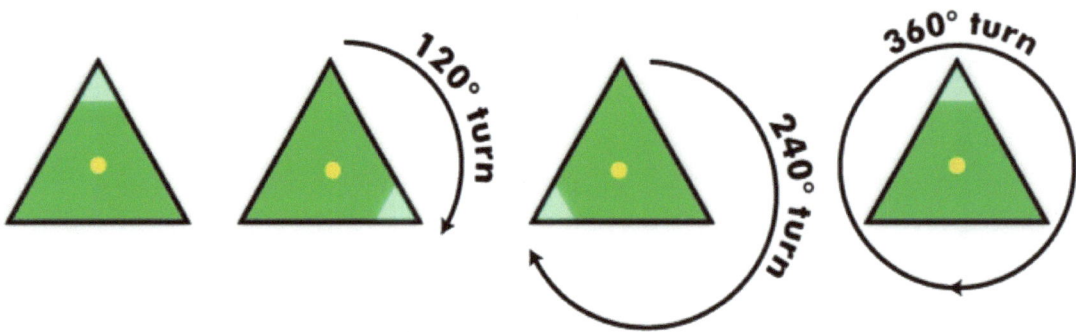

120° and 240°. At 360° the triangle has returned to its original position.

Circle the figures that have rotational symmetry.

Complete these shapes so that they have rotational symmetry.

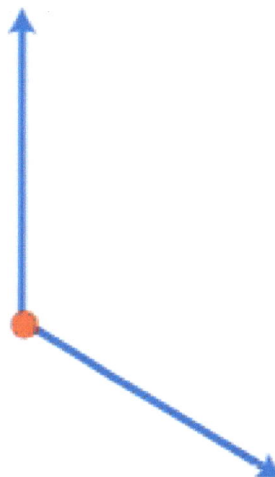

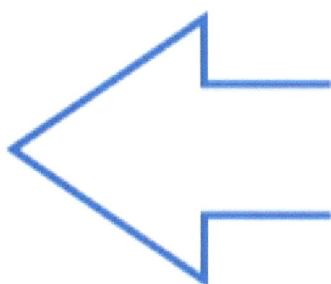

Name: _____

Exploring Symmetry Quiz

1 True or false? This figure has rotational symmetry.

2 Which shape has no lines of symmetry?

A　　B　　C　　D

3 How many lines of symmetry does this shape have?

4 How many lines of symmetry does this shape have?